LINCOLN MOTOR CARS
1946 THROUGH 1960
PHOTO ARCHIVE

LINCOLN MOTOR CARS 1946 THROUGH 1960 PHOTO ARCHIVE

Photographs from the Detroit Public Library's National Automotive History Collection

Edited with introduction by Mark A. Patrick
Curator, National Automotive History Collection

Iconografix
Photo Archive Series

Iconografix
PO Box 609
Osceola, Wisconsin 54020 USA

Library of Congress Card Number 96-76228

ISBN 1-882256-58-1

96 97 98 99 00 01 02 03 5 4 3 2 1

Digital imaging by Pixelperfect, Madison, Wisconsin
Cover design by Lou Gordon, Osceola, Wisconsin

Printed in the United States of America

US book trade distribution by Voyageur Press, Inc. (800) 888-9653

PREFACE

The histories of machines and mechanical gadgets are contained in the books, journals, correspondence, and personal papers stored in libraries and archives throughout the world. Written in tens of languages, covering thousands of subjects, the stories are recorded in millions of words.

Words are powerful. Yet, the impact of a single image, a photograph or an illustration, often relates more than dozens of pages of text. Fortunately, many of the libraries and archives that house the words also preserve the images.

In the *Photo Archive Series,* Iconografix reproduces photographs and illustrations selected from public and private collections. The images are chosen to tell a story—to capture the character of their subject. Reproduced as found, they are accompanied by the captions made available by the archive.

The Iconografix *Photo Archive Series* is dedicated to young and old alike, the enthusiast, the collector and anyone who, like us, is fascinated by "things" mechanical.

A new side nameplate, heavier bumper and egg crate grille distinguished the 1946 "warmed-over" pre-war Lincoln.

INTRODUCTION

The story of Lincoln motor cars, in the period following World War II, was one of resounding success. Unquestionably, Lincoln's success paralleled the broad advances of the American economy. However, the fact that Lincoln remained a symbol of American superiority is contrasted by a sense of drama. In 1942, the future of the Lincoln division was not so bright.

Sales of Lincoln hit a record 30,974 units in 1937 only to drop to 19,527 for1938; 21,133 in 1939; 22,046 in 1940; 20,094 in 1941; and 6,547 in the abbreviated model year of 1942. The conversion to wartime production further unsettled the picture, then came the untimely death of Edsel Ford in the spring of 1943. Henry Ford was disabled by a series of strokes during the war. These unfortunate developments led to power struggles and rivalries within the company. It was not a healthy situation. The government was sufficiently concerned about the viability of Ford Motor Company to dispatch Naval Second Lieutenant Henry Ford II, grandson of the founder, back to Dearborn, with the mission of putting things back on course. It wasn't going to be easy, however. Clara Ford, his grandmother, and Eleanor, his mother, two of the most respected women associated with the industry, had to navigate corporate intrigue to assist young Ford in seizing back the reins of the family company. Their rival, Harry Bennett, showed signs of leading a corporate putsch. This led to a dramatic showdown between Ford and Bennett, at which Bennett was dismissed from employment.

At this stage, the Truman administration weighed in with an important decision. Victory over the Axis powers was certain by late 1944. The demilitarization of the American economy was a sensitive political issue, especially following on the heels of the Great Depression. Economic nabobs commanded that Ford be the first automotive company to resume domestic automobile production—even as war waged on in the Pacific. This, of course, fueled resentment within the industry, but it was widely recognized that Ford was on shaky ground and needed to be saved at all costs.

Although production of Fords began on July 3, 1945, production of Lincolns was delayed until early 1946. When the two 1946 models finally arrived, they were essentially identical to the pre-war cars. The Zephyr name was dropped, the car was referred to simply as Lincoln, but the name Continental was carried on. The Lincolns of first three model years following the war varied little from each other. The photographs which follow document very well the slight variations between the models.

The 1949 model year brought the first true post-war Lincolns. Gone were the Continental, replaced by a new luxury model, the Cosmopolitan, and the Lincoln V-12, which was replaced by a powerful 152 bhp V-8. Lincolns took on a more modern style but retained suicide door. The grille lost most of its vertical lines.

The standard Lincoln and Cosmopolitan changed little for 1950 and 1951, but were completely redesigned for 1952. The Lincoln took on a more svelte appearance.

The grille all but disappeared into a massive chrome bumper. The Cosmopolitan was down-graded to the base model and the new Capri became the luxury model. The Lincoln models for 1953, 1954, and 1955 were all similar to the 1952 series, with a boost of horsepower in 1953 and 1955, and annual variations in chrome trim and the like.

The next major restlying occurred in 1956. A faint hint of a grille reappeared above and below the broad horizontal bumper. Hooded headlights added a touch of elegance to styling that contrasted nicely with the cars' overall sporty design. The Capri now became the base model, with the introduction of the new luxury Premiere. These cars have a decidedly modern look about them, and are considered by many to be the most beautiful post-war Lincolns.

Further adding to the success of the 1956 model year was the reintroduction of the Continental. Unquestionably one of the most important post-war automobiles, the Continental Mark II was distinguished by its sleek and lengthy hood design and its understated use of chrome and delicate grille work. Its modern appearance gave it the essence of an experimental car. Like many classic cars of the 1930s, the Mark II was associated with successful and beautiful actors, business executives, and sports stars. The Mark II was the automotive designers way of concluding the grim McCarthy era—it was extremely refreshing.

Although the Continental Mk. II remained virtually unchanged for 1957, its final year of production, both Capri and Premiere models followed the crowd and sprouted tailfins. Although the fins were shed for 1958, the remaining distinguishing feature of the 1957 model year, Quadra-Lites, pairs of separate low and high beams, remained. The Capri and Premiere were completely restyled for 1958. Their roofline, with its slanted rear quarter, easily distinguished these new models. Although offered in four body styles including a convertible, Continental lost all of its individuality in 1958, becoming little more than a Premiere bearing the name Continental Mk. III.

For 1959 and 1960, the final years covered by this work, the Capri and Premiere changed little from 1958. The pathetic Continental gained little other than added Roman numerals, becoming the Mark IV in 1959; the Mark V in 1960.

Lincoln Motor Cars 1946 through 1960 Photo Archive and its companion *Lincoln Motor Cars 1920 through 1942 Photo Archive* celebrate one of the remaining titans of the auto industry. With each, the reader can expect a glorious photographic history of some of the greatest American automobiles ever produced. The photographs are a part of the Detroit Public Library's National Automotive History Collection. The mission of the NAHC is to retain and preserve the historical record of the automobile and other forms of wheeled transportation. Toward this aim, NAHC has become the premier collection of its type. Our files include 350,000 pieces of sales literature and 250,000 photographs. The collection also houses biographical files, books, magazines, art, blueprints, owner's manuals, and personal papers of automotive pioneers and trailblazers. Much of this material is unique. Most importantly, it is in the public library, and, therefore, is accessible to the enthusiast.

I offer special thanks to Michael Lamm, David R. Holls, Byron D. Olsen, and NAHC Librarian Serena Gomez, for their assistance in identification and for their useful recommendations.

Lincoln 1946 through 1948

1946 Model 73 Six-Passenger Sedans await grilles and bumpers, due to early postwar shortages of chromium.

Adding brightwork during the final assembly stages of 1946 Model 73 Six-Passenger Sedans, at left, and 1946 Model 77 Six-Passenger Club Coupes, at right.

These two views of the 1946 Model 73 Six-Passenger Sedan illustrate the minimal changes made to this still-handsome pre-war design.

Front view of the 1946 Model 73 Six-Passenger Sedan.

The front interior of the 1946 Model 73 Six-Passenger Sedan.

A 1946 Model 76 Six-Passenger Convertible.

The 1947 and 1948 Lincolns are virtually indistinguishable from one another. Yet, they are easily distinguished from 1946 models (see inset): pull-type door handles replaced push buttons; a more tasteful Lincoln script replaced the heavy Lincoln name plate on the hood sides; hub caps were flatter, with the hexagonal raised center eliminated. This is a Model 73 Six-Passenger Sedan.

17

Lincoln 1949 through 1951

The 1949 Cosmopolitan Series was top-of-the-line for Lincoln. This is a Model 9EH Club Coupe.

These Lincoln Motor Company Engineering photos accurately convey the girth of the 1949 Model 9EH Sport Sedan.

Rear view of the 1949 Cosmopoolitan Model 9EH Sport Sedan.

Rear view of the 1949 Cosmopolitan Model 9EH Town Sedan.

Two 1949 Model9EL Sport Sedans with elegant two-toned paint schemes.

The 1949 Model 9EH Cosmopolitan Four-Door Town Sedan. The Lincoln-Mercury News Bureau trumpeted its, "streamlined elegance and maximum comfort."

Interior views of the 1949 Model 9EH Cosmopolitan Four-Door Town Sedan. Power windows and seats were standard.

Model 9EH Cosmopolitan
Four-Door Town Sedan rear
seating was reminiscent of
pre-war formal sedans.

Painted in a lighter color scheme, the 1949 Model 9EH Cosmopolitan Four-Door Town Sedan had a decidedly less-formal appearance.

Long, low lines characterized the fleetness of the 1949 Model 9EL Convertible.

Serving the White House and US Presidents, this 1950 custom limousine was based on a Model OEH Cosmopolitan Convertible. The removable 'bubble-top' was added in 1954.

A 1950 Model OEL Club Coupe.

Introduced mid-year, the elegant 1950 Model OEL Lido Coupe featured a vinyl top and custom interior.

A 1951 Model OEH Cosmopolitan Four-Door Sport Sedan.

A 1951 Model OEL Four-Door Sport Sedan.

Two views of the 1951 Model OEH Cosmopolitan Convertible.

Lincoln 1952 through 1955

The 1952 Model 73A Cosmopolitan Four-Door Sedan.

A top-of-the-line 1952 Model 60A Capri Special Custom Coupe.

A 1952 Model 60C Cosmopolitan Custom Sport Coupe.

A 1953 Model 60A Capri Hardtop Coupe.

A 1953 Model 73B Capri Four-Door Sedan. The gold-plated "V" on the rear door commemorated the 50th anniversary of Ford Motor Company.

A 1953 Model 76A Capri Convertible. The Capri carried a gold-plated "V" above the front bumper.

A 1953 Model 73A Cosmopolitan Four-Door Sedan.

Rear interior and dashboard of the 1953 Model 73A Cosmopolitan Four-Door Sedan.

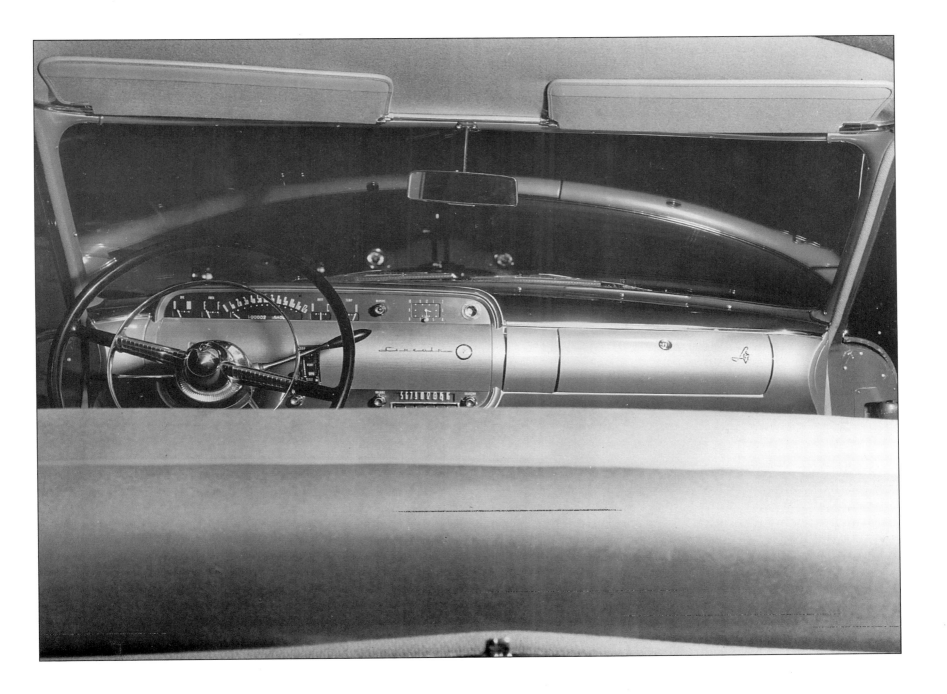

For 1953, Lincoln offered a new 205-horsepower V-8 as standard equipment. Power brakes, power steering, and four-way power seat were optional.

A 1954 Model 73A Cosmopolitan Four-Door Sedan. Lincolns were longer and wider in 1954.

Do come in. The interior of a 1954 Model 73A Cosmopolitan Four-Door Sedan.

Dashboard of a 1954 Model 73A Cosmopolitan Four-Door Sedan.

A General Motors Proving Ground's photo of the engine compartment of the 1954 Lincoln Cosmopolitan Sedan.

Two views of the 1954 Model 73B Capri Four-Door Sedan.

Interior views of the 1954 Model 73B Capri Four-Door Sedan.

Dashboard of a 1954 Model 73B Capri Four-Door Sedan.

A 1954 Model 76A Capri Convertible.

A 1954 Model 60A Capri Hardtop Coupe.

A 1955 Model 60A Capri Hardtop Coupe. Obvious differences from the 1954 model (opposite) include taillight treatment and chrome rocker panels.

Front view of the Model 60A Capri Hardtop Coupe displays a new hood ornament for 1955.

Optional continental kit fitted to a 1955 Model 76A Capri Convertible.

1955 Model 73B Capri Four-Door Sedan. Nameplate, chrome rocker panels, and different rear roof pillars distinguished the Capri from the new lower-priced Custom (the Cosmopolitan was dropped after 1954). Air intake vents on rear fender and ducts visible through rear window indicate trunk-mounted factory air conditioning.

Lincoln 1956 through 1957

Clay mockup of the 1956 Model 73A Capri Four-Door Sedan.

Production Model 73A Capri Four-Door Sedan. Lincoln's publicity department touted Lincoln's "big new look". The 1956 models were more than 18-1/2-feet long, seven-inches longer and 2-1/2-inches lower than the 1955 model.

Clay mockup of a car that did not reach production, the 1956 Capri Convertible.

The "big new look" was obvious here. A 1956 Model 73B Premiere Four-Door Sedan with optional two-tone paint.

Lincoln's sole convertible for 1956, the Model 76B Premiere. Standard Premiere features included automatic transmission, power steering, windows, and four-way seat.

The 1956 Premiere Convertible interior offered two-tone leather seats and trim.

Safety devices figured prominently for Lincoln in 1956: safety-flex type steering wheel, with 3-1/2-inches of energy-absorbing "cushion"; seat belts for driver and passenger; double-grip rotor-type door locks; glare-reducing textured-finish vinyl covering for the instrument panel; and vinyl backed rear-view mirrors to minimize shattering.

1957 brought "Quadra-Lites" and tailfins to Lincoln.

1957 also brought the "pillarless" look to the four-door lineup. A Model 57B Premiere Four-Door Landau Hardtop Sedan, pictured with a 1957 Continental Mk. II.

The 1957 Premiere, except for nameplate and identifying script on the front fender, was virtually indistinguishable from the lower-priced Capri.

Ease of entry, an unobstructed view, and, of course, the beauty of a hardtop were the features Lincoln promoted with photos such as these.

A 1957 Model 58A Capri "Concealed-Pillar" Four-Door Sedan. It used a slim pillar in construction, hidden when the windows were raised.

A 1957 Model 60B Premiere Hardtop Coupe.

Big but beautiful, a 1957 Model 76B Capri Convertible and (opposite) Capri Convertible interior.

Lincoln 1958 through 1960

Completely restyled for 1958, the Model 57A Four-Door Landau Hardtop Sedan.

A Model 63B Premiere Two-Door Hardtop. Lincoln for 1958 was five-inches longer than in 1957, and, for the first time, built employing unitized construction.

Rear view of the 1958 63B Premiere Two-Door Hardtop. The chrome rocker panel, side nameplate, and star on the front door distinguished the Premiere from the lower-priced Capri.

The 1958 Lincolns featured a new 375 bhp 430-cubic inch V-8 with Holley four-barrel carburetor. The car is a Model 53A Capri Four-Door Sedan.

Luggage compartment of the 1958 Model 53A Capri Four-Door Sedan.

Two views of a 1959 Model 57B Premiere Hardtop Sedan. The 1959 Premiere was offered in four-door sedan, four-door hartop ("landau"), and two-door hartop versions.

The 1959 Model 57B Premiere Hardtop Sedan. Two inches shorter than the 1958 model, it did not appear so.

Instrument panel of the 1959 Model 57B Premiere.

Two views of the 1960 Model 57A Standard Four-Door Hardtop Sedan.

A 1960 Model 57B Premiere Four-Door Hardtop Sedan. The Premiere was distinguished by the small medallion behind the front wheel well.

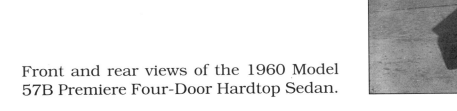

Front and rear views of the 1960 Model
57B Premiere Four-Door Hardtop Sedan.

Continental Mk. I 1946 through 1948
Continental Mk. II - Mk IV 1956 through 1960

Raymond Loewy-designed and Derham-built custom 1946 Continental Coupe—in British parlance a Sedanca Coupe.

A 1946 Model 66H/57 Continental Two-Door Club Coupe.

Difficult to distinguish from 1946 or 1948, this is a 1947 Model 76H/52 Continental Convertible.

A 1948 Model 876H/56 Continental Convertible.

Another 1948 Model 876H/56 Continental Convertible with top up and retracted.

Two views of the clay mockup for the revived Continental, the 1956 Mk. II.

The production 1956 Mk. II Continental, photographed in front of the new Continental Division headquarters.

Simple, clean, and elegant, the front end of the 1956 Mk. II Continental.

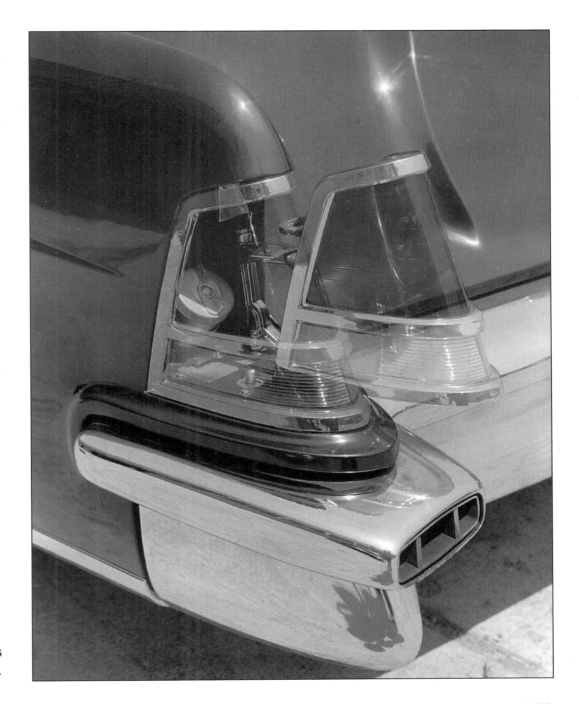

The gasoline filler cap was
hidden behind the taillight.

Interior of the 1956 Mk. II. Everything but air conditioning was standard—leather, full-power options, radio, etc.

A 1957 MK. II Continental Sport Coupe, virtually identical to the 1956 version.

One of only two 1957 Mk. II Continental Convertibles.

Certainly less distinguished than the Mk. II, but a Continental nonetheless, the 1958 Mk. III. This was the Model 75A Four-Door Hardtop Sedan, one of four Continental models offered.

A 1958 Model 65A Mk. III Continental Two-Door Hardtop Coupe. Note the retractable rear window.

Roof line of the 1958 Model 68A Mk. III Continental Convertible was common to that of the rest of the line. These photos show the roof retracting into the rear storage compartment.

The 1958 Model 68A Mk. III Continental Convertible with roof fully-retracted.

Two views of the 1959 Model 75A Mk. IV Continental Four-Door Hardtop Sedan.

118

Interior of the 1959 Model 75A Mk. IV Continental Four-Door Hardtop Sedan. The Mk. IV was equipped with six-way power seats, tinted glass, radio with dual-speakers, and power windows (including vent).

A 1959 Model 23A Mk. IV Continental Six-Passenger Executive Limousine. Sold in black only, many metal surfaces were gold finished. A retracting glass partition separated the chauffeur from the elite.

A new grille and new side chrome were among the few distinguishing characteristics of the 1960 Mk. V. This is the Model 75A Four-Door Hardtop Sedan.

A 1960 Model 65A Mk. V Continental Two-Door Hardtop Coupe.

The 1960 Mk. V offered a completely new instrument panel cluster featuring four hooded circular pods, housing gauges and indicators, set above the panel face. Chrome plating and padding galore!

Round taillights and a similar pattern to that of the front grille distinguished the 1960 Mk. V rear end.

The Iconografix Photo Archive Series includes:

The Iconografix Photo Archive Series is available from direct mail specialty book dealers and bookstores worldwide, or can be ordered from the publisher. For additional information or to add your name to our mailing list contact:

Iconografix
PO Box 609/BK
Osceola, Wisconsin 54020 USA

Telephone: (715) 294-2792
(800) 289-3504 (USA)
Fax: (715) 294-3414

Book trade distribution by Voyageur Press, Inc., PO Box 338, Stillwater, Minnesota 55082 USA (800) 888-9653
European distribution by Midland Publishing Limited, 24 The Hollow, Earl Shilton, Leicester LE9 7N1 England

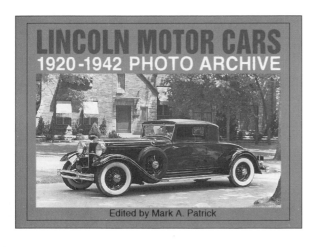

MORE GREAT BOOKS FROM ICONOGRAFIX

LINCOLN MOTOR CARS 1920-1942 *Photo Archive* ISBN 1-882256-57-3

PACKARD MOTOR CARS 1946-1958 *Photo Archive* ISBN 1-882256-45-X

IMPERIAL 1955-1963 *Photo Archive* ISBN 1-882256-22-0

STUDEBAKER 1946-1958 *Photo Archive* ISBN 1-882256-25-5

DODGE TRUCKS 1948-1960 *Photo Archive* ISBN 1-882256-37-9

COCA-COLA: A HISTORY IN PHOTOGRAPHS 1930-1969 *Photo Archive* ISBN 1-882256-46-8

AMERICAN SERVICE STATIONS 1935-1943 *Photo Archive* ISBN 1-882256-27-1

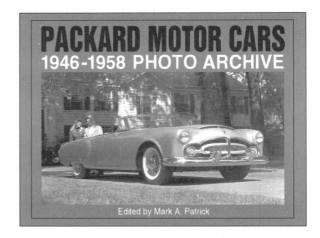

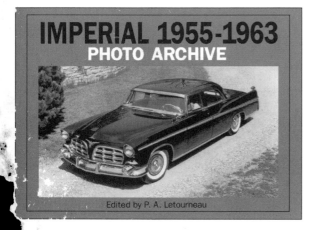

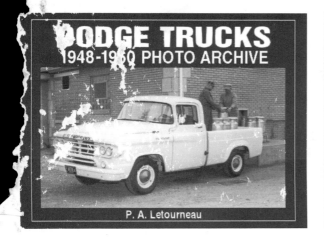

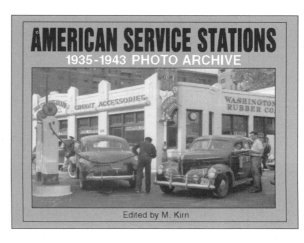